ELEMENTS OF THE PERIODIC TABLE

WHAT IS HELIUM?

002

He

HELIUM

4.00

KATHLEEN A. KLATTE

ROSEN PUBLISHING

Published in 2026 by The Rosen Publishing Group, Inc.
2544 Clinton Street, Buffalo, NY 14224

Portions of this work were originally authored by
Heather Hasan and Jun Lim and published as *Helium*. All new
material this edition authored by Kathleen A. Klatte.

Designer: Rachel Rising
Editor: Kathleen A. Klatte

Library of Congress Cataloging-in-Publication Data
Names: Klatte, Kathleen A, author.
Title: What is helium? / Kathleen A Klatte.
Description: Buffalo, New York : Rosen Publishing, [2026] | Series:
Elements of the periodic table | Includes bibliographical references and
index.
Identifiers: LCCN 2024061179 | ISBN 9781499478907 (library binding) | ISBN
9781499478891 (paperback) | ISBN 9781499478914 (ebook)
Subjects: LCSH: Helium--Juvenile literature. | Chemical elements--Juvenile
literature.
Classification: LCC QD181.H4 K53 2026 | DDC 546/.751--dc23/eng/20250212
LC record available at https://lccn.loc.gov/2024061179

Manufactured in the United States of America

CPSIA Compliance Information: Batch #CSRYA26. For further information, contact
Rosen Publishing at 1-800-237-9932.

CONTENTS

INTRODUCTION

Helium (He) is the second-lightest element in the known universe. It's colorless, odorless, tasteless, and nonflammable. Had it been used to fill airships in the early 20th century, some major aeronautic disasters might have been avoided.

An airship or dirigible is a lighter-than-air, self-propelled aircraft. Unlike hot-air balloons that drift with the air currents, dirigibles can be steered. The balloon part of early 20th century dirigibles was designed to be filled with helium or hydrogen (H). Helium is the much safer choice, as it doesn't burn.

So, if helium is safer, why was hydrogen used at all? The answer is primarily economic reasons. Hydrogen is the lightest element. In practical terms, this means it has greater lifting power than helium. An airship filled with hydrogen could lift greater payloads than one filled with helium. This means it could carry more cargo or passengers. A ship using hydrogen for lift can also travel farther without having to refuel. In the years leading up to World War II, helium was also hard to come by. This led to airships using hydrogen for lifting gas.

However, as history has shown, this was a dangerous trade. The U.S. military stopped using hydrogen to fuel its airships in 1922, after the airship named *Roma* crashed. The *Hindenburg* disaster of 1937 is attributed to a spark which ignited the hydrogen and destroyed the airship in 32 seconds. Although recent technological breakthroughs have made it possible to fuel modern airships safely with hydrogen, some now use helium for lift.

So, what else do we use helium for today? Helium is used for certain types of welding. It's also mixed with oxygen for pressurized breathing, such as in scuba gear. Helium has a freezing point so low that it can't be solidified naturally. Liquid helium is so cold that it's used as a refrigerant in cryogenics. Of course, its most familiar use is in balloons. It's used in both colorful birthday balloons and balloons used by the military and for weather research.

Although you're probably most familiar with helium-filled party balloons, the second lightest- element in the universe has many more uses, especially in science and medicine.

CHAPTER ONE

GETTING TO KNOW HELIUM

Helium is something of an enigma. It's the second most common element in the universe, yet it makes up only 0.0005 percent of Earth's atmosphere. It's the second-lightest element and the coldest. There's so little helium in our atmosphere because Earth's gravity isn't strong enough to hold it. Its natural phase is gaseous, and it must be cooled to temperatures of −452°F (−268.9°C) to liquify. Helium doesn't occur naturally as a solid.

Helium is a noble gas, an element that's part of group 18 of the periodic table. The noble gases are generally considered inert, or nonreactive. This means they don't usually bond with other elements. However, unlike truly inert gases, the noble gases can bond with other elements under certain conditions.

So, if there's so little helium in the air, and it's not found in solid form, where is it? Most helium is found in space. Stars, including our sun, are huge balls of burning gas, mainly hydrogen. In the star's core, the process of nuclear fusion transforms hydrogen into helium. This is what produces the light and heat that make life on Earth possible.

Eventually, the heat and pressure inside the star will convert the helium into carbon. This process takes many centuries.

NASA scientists studying data from space telescopes think there might be planets in other solar systems with helium in their atmosphere. Scientists think these planets are similar to Neptune but smaller and closer to their sun. This would make them hot enough to turn the hydrogen in their atmosphere to helium.

Hydrogen turns into helium through a process called fusion. The only place in nature hot enough to fuse two protons together is inside a star. This diagram shows the inside of the sun, which is a star, with temperatures measured in kelvin.

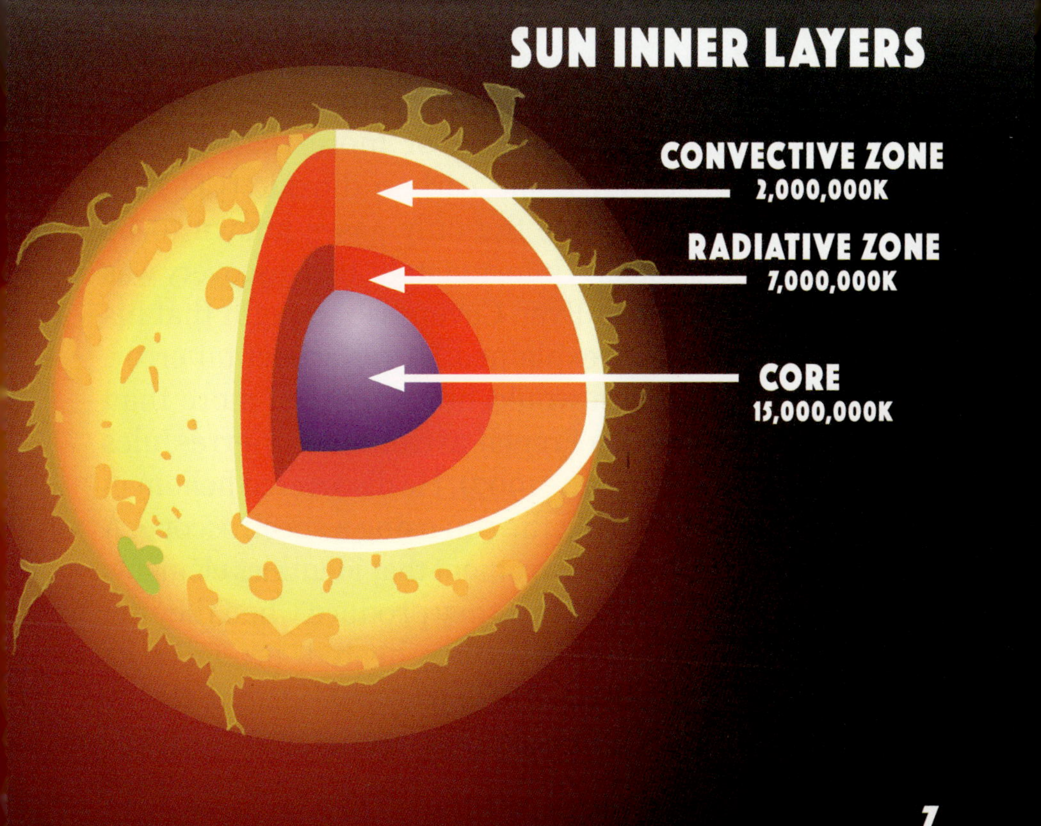

On Earth, nearly all helium is found trapped underground. In fact, 3,000 times more helium is found trapped below the surface of Earth than is found in Earth's atmosphere. Much of the helium that is found underground is in reservoirs of natural gas. Natural gas is a mixture of gases trapped in the pores of underground rocks. Helium forms underground when radioactive, or unstable, elements in these rocks decay, or break down. When elements such as uranium (U) and radon (Rn) decay, they spontaneously give off helium. Some of this helium seeps up through cracks in the rocks and escapes into the atmosphere, while some remains trapped within the rocks.

WHO DISCOVERED HELIUM?

In 1785, a man named Henry Cavendish (1731–1810) found that air was made of more than oxygen (O_2), nitrogen (N_2), and carbon dioxide (CO_2). At that time, they were the only gases known to be components of air. Unfortunately, this discovery was largely forgotten. Even Cavendish thought that his results could be explained by experimental error. However, helium would be discovered more than 80 years later by French astronomer Pierre-Jules César Janssen (1824–1907). He discovered helium in 1868 while studying the sun using a spectroscope. A spectroscope is an instrument that identifies elements by the spectrum of lines they produce when they are heated. Each element, when heated, produces its own spectrum of colored lines. While examining the spectrum from the sun, Janssen noticed a yellow line that did not belong to any element known at that time. Meanwhile, British astronomer J. Norman Lockyer (1836–1920) also observed the mysterious yellow line. Lockyer concluded that it belonged to an element that was common in the sun but was unknown on Earth. Therefore, he named the element "helium" after Helios, a Greek god of the sun. Many scientists doubted the existence of helium. Even so, a search for the new element on planet Earth began.

The planet Neptune in our solar system is what's called an ice giant. It appears blue due to methane in its atmosphere. NASA scientists think there could be planets like Neptune, but warmer, with helium in their atmospheres, in other solar systems.

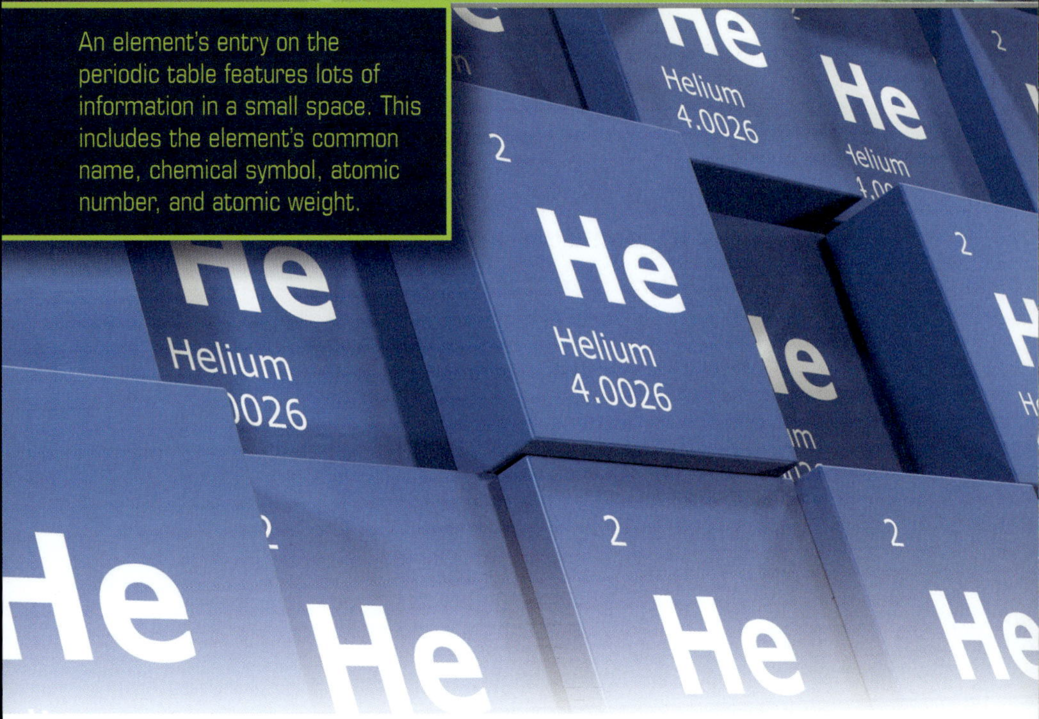

An element's entry on the periodic table features lots of information in a small space. This includes the element's common name, chemical symbol, atomic number, and atomic weight.

The hunt to find helium on Earth finally ended in 1895. William Ramsay (1852–1916), a Scottish chemist, discovered helium on Earth while experimenting with cleveite, a uranium-containing mineral. A mineral is a crystalline substance of definite composition that occurs naturally in rocks. Ramsay exposed the cleveite to mineral acids and collected the gases that were produced. He sent a sample of the gases to Lockyer and William Crookes (1832–1919), a British physicist. They were able to identify helium as one of the gases by using the same spectroscopic method used to identify helium in the sun. Two Swedish chemists, Nils Abraham Langlet (1868–1936) and Per Teodor Cleve (1840–1905), also independently discovered helium in cleveite at about the same time as Ramsay. Helium is the only element to be identified in space before being found on Earth!

HELIUM SNAPSHOT

CHEMICAL SYMBOL: He

CLASSIFICATION: Nonmetal

PROPERTIES: Nonreactive, nonflammable, colorless, odorless, nontoxic

DISCOVERED BY: Pierre-Jules César Janssen, in 1868, while studying the sun using a spectroscope. Helium was discovered on Earth by William Ramsay, Nils Abraham Langlet, and Per Teodor Cleve in 1895.

ATOMIC NUMBER: 2
ATOMIC WEIGHT (ROUNDED):
4.0026 atomic mass units (amu)

PROTONS: 2 **ELECTRONS:** 2 **NEUTRONS:** 2

DENSITY AT 68°F (20°C): 0.0001664 gram/cubic centimeter (0.1664 g/liter)

MELTING POINT: -457.87°F (-272.15°C)

BOILING POINT: -452.09°F (-268.94°C)

COMMONLY FOUND: Air, stars, underground

2 4.0026
He
Helium

An element's atomic number is the number of protons in its nucleus. Helium's atomic number is two.

Dmitri Mendeleev published the first version of the periodic table in 1869. Today the table is maintained and published by the International Union of Pure and Applied Chemistry (IUPAC). They approve the addition of newly discovered elements.

MENDELEEV'S TABLE

Today, the periodic table of elements is one of the most easily recognized images in the world. It's posted in labs and classrooms around the globe and reprinted in a multitude of textbooks. But it wasn't always that way. In the mid-19th century, a respected Russian chemist named Dmitri Mendeleev (1834–1907) was writing a chemistry textbook. He realized there wasn't a good method of organizing the known chemical elements for study and decided to do something about it.

Mendeleev organized all the known elements in a table according to their atomic weight. When the elements were arranged this way, he could see patterns between substances with similar atomic weights. Mendeleev lived at a time when many great discoveries were being made about the natural world. He believed there were still more elements to be discovered and that they would fill in gaps in his table. His theory was proven correct, with additional elements—including helium—being discovered and added to the table within his lifetime.

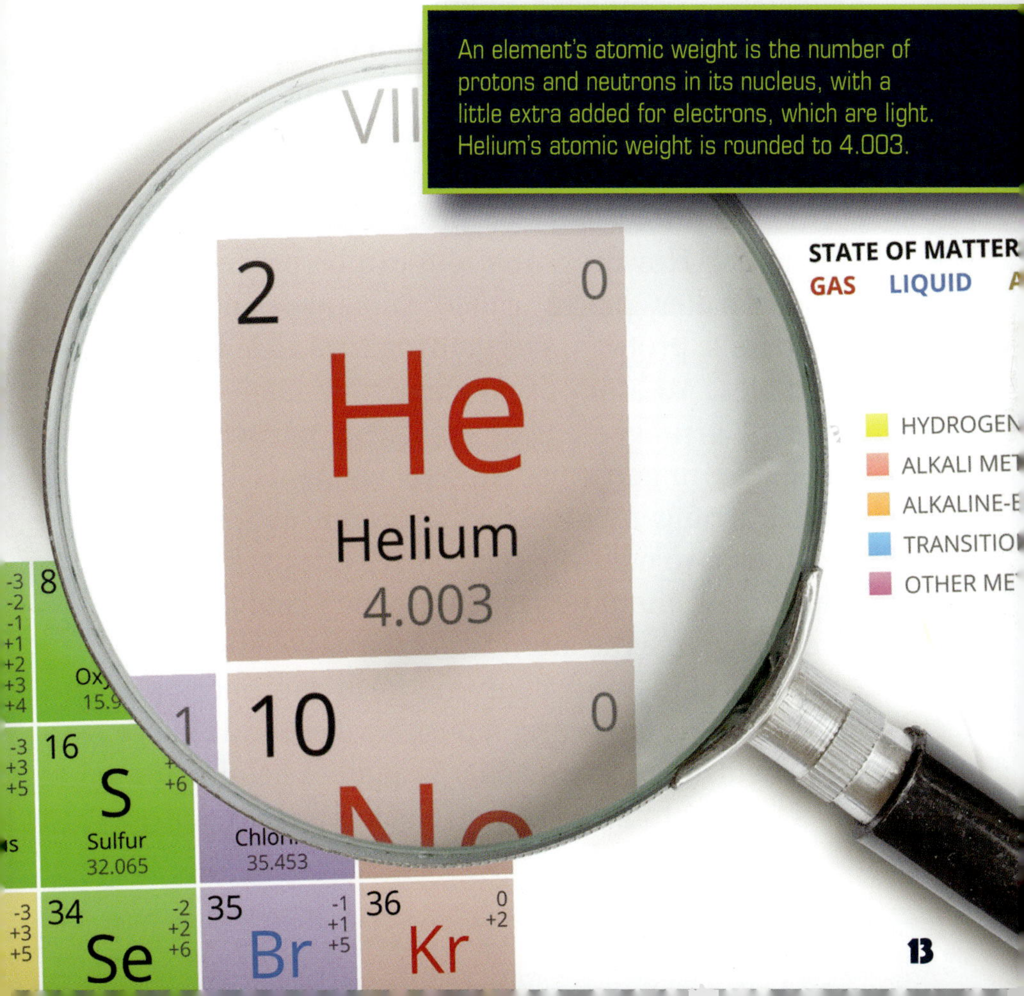

An element's atomic weight is the number of protons and neutrons in its nucleus, with a little extra added for electrons, which are light. Helium's atomic weight is rounded to 4.003.

CHAPTER TWO

WHAT MAKES HELIUM SPECIAL?

Helium is one of 118 chemical elements currently included on the periodic table. An element is a substance composed of only one kind of atom. Unlike compounds such as water (H_2O), which can be broken into two hydrogen atoms and one oxygen atom, helium is only made up of helium atoms.

An atom is the smallest particle of an element. A helium atom can't be broken down into other elements. However, an atom can be broken down into smaller pieces called subatomic particles. Electrons, protons, and neutrons are all subatomic particles. An atom of an element is made up of a specific arrangement of subatomic particles—sort of like a fingerprint. No two elements have the same arrangement. A helium atom contains two protons, two neutrons, and two electrons. Changing the arrangement of particles (which isn't possible) would change the atom to a different element.

Protons and neutrons form the nucleus of an atom. Protons carry a positive electrical charge, and neutrons carry no charge at all. A helium atom has two protons, so its nucleus has a charge of +2.

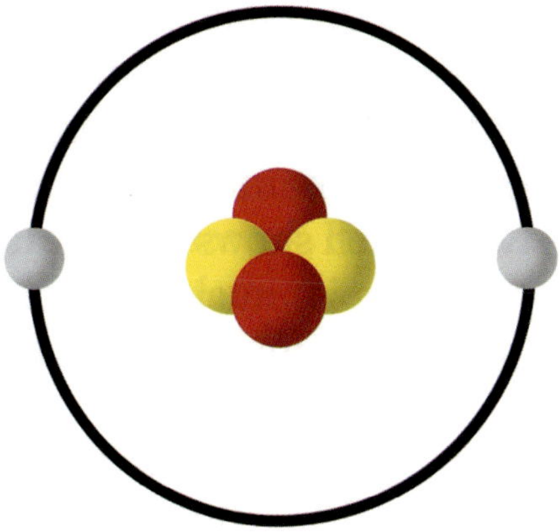

Helium has two protons in its nucleus. This means its atomic number is two. The atomic mass of helium is 4.

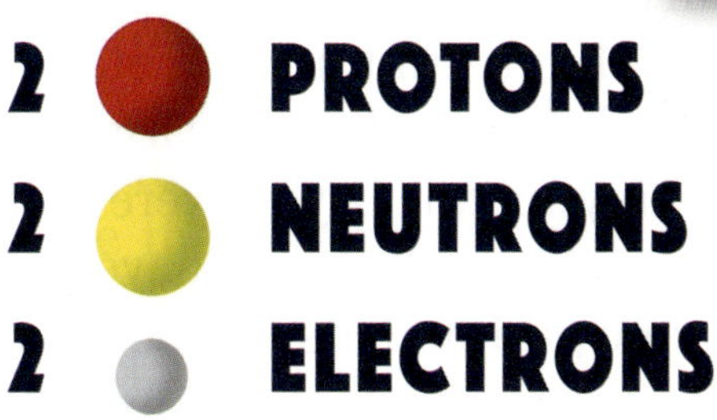

Electrons orbit, or circle, around the nucleus. They have a negative charge. The attraction between the positive and negative charges are what holds the atom together. Atoms have the same number of electrons and protons.

ELEMENTAL DIFFERENCES

What makes helium different from other elements, such as oxygen or silver (Ag)? The difference lies in the number of protons in the nucleus of an atom. The number of protons in an atom's nucleus determines which element it is. This number is called the

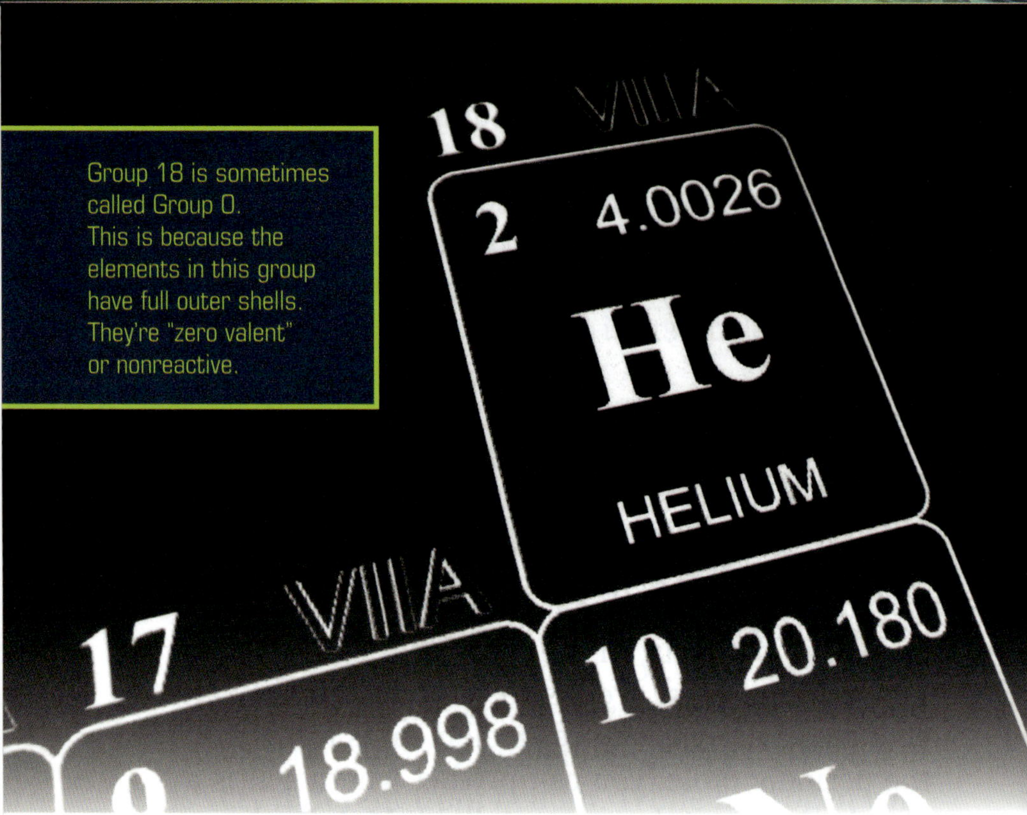

Group 18 is sometimes called Group 0. This is because the elements in this group have full outer shells. They're "zero valent" or nonreactive.

can be pulled into wires. Substances are classified as nonmetals if they lack the characteristics of metals. Nearly half of the nonmetals are colorless gases, but nonmetals can also be liquids and solids. Unlike solid metals, however, solid nonmetals are brittle and will crumble or break apart if pulled.

As you look across the periodic table from left to right, a horizontal row of elements is called a period. Elements are arranged in periods by the elements' atomic numbers. From an element's location on the periodic table, a person can determine the arrangement of the electrons in an atom. Helium is in period one, meaning it has only one shell of electrons surrounding its nucleus.

As you read down the periodic table from top to bottom, a column of elements is called a group or family. Just as you might have similar characteristics to the other members of your family, the elements in any given group have similar properties. Helium heads the family of elements that make up group O, or 18. This group of elements is also referred to as the noble gases. The noble gas family has six gaseous elements: helium (He), neon (Ne), argon (Ar), krypton (Kr), xenon (Xe), and radon (Rn).

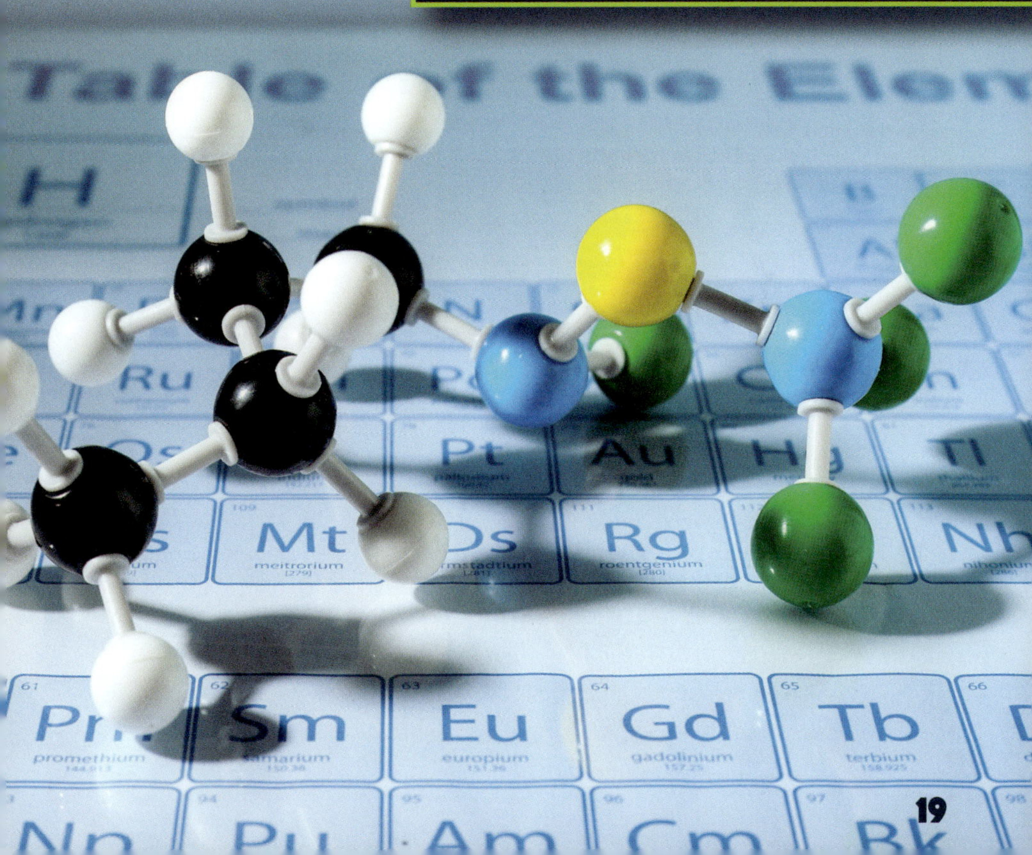

Noble gases are naturally nonreactive, but scientists have been able to create chemical reactions with them under laboratory conditions.

GROUP 18

Group 18 of the periodic table of elements has several colorless, odorless gases, including helium. These are called the noble gases. Initially, these gases were believed to be inert—meaning they could not form compounds with other elements. However, new technology has revealed that the noble gases can combine with other elements under certain conditions.

NOBLE GAS	INERT GAS
ELEMENT IN GROUP 18 OF THE PERIODIC TABLE	COMPOUND THAT DOES NOT REACT WITH OTHER ELEMENTS UNDER NORMAL CONDITIONS
COMPOSED OF ONLY ATOMS (ELEMENTAL)	COMPOSED OF ATOMS OR MOLECULES
CAN REACT WITH OTHER ELEMENTS UNDER CERTAIN CONDITIONS	CANNOT REACT CHEMICALLY WITH OTHER ELEMENTS
INCLUDES THE ELEMENTS IN GROUP 18	INCLUDES THE NOBLE GASES PLUS ADDITIONAL GASES

A chemical reaction occurs when two atoms share the molecules from their outer, or valence, shell. For example, a molecule of water is formed when two hydrogen atoms and one oxygen atom share electrons, forming a chemical bond. Inert gases can't form bonds because their valence shells are already full of electrons. There's no room to add more. Atoms with full outer shells are considered stable.

Recently, element 118, oganesson (Og), has been included in group 18 as a noble gas. Oganesson is the one of the most recent additions to the periodic table. It was added in 2016.

The term "noble" dates to the medieval practice of alchemy. Some of the precious metals, such as gold, are also referred to as noble. The noble metals are a group of metallic elements that are generally not reactive to oxygen.

THERMONUCLEAR EXPLOSIONS

A thermonuclear device or hydrogen bomb operates by fusing hydrogen atoms to form helium. When two hydrogen atoms combine, some of their mass is forced out of the new helium molecule. This force is converted to energy. The energy released by this chemical reaction is what causes the immense destruction of the bomb. When this kind of bomb, called an H-bomb, detonates, it creates a self-sustaining chain reaction of hydrogen molecules fusing into helium. Temperatures of over a million degrees are required to start the reaction. H-bombs have been detonated in tests with extraordinarily destructive results, vaporizing entire islands. They're exponentially more destructive than the atomic bombs used in World War II. H-bombs have never been used in combat.

Atomic bombs, such as the ones detonated in Japan at the end of World War II, function by splitting atoms. This process is called fission. In addition to the immediate destruction caused by the blast and shockwave of the explosion, both atomic bombs and hydrogen bombs produce radioactive fallout that contaminates the bomb site for generations. The area around the islands where thermonuclear devices were tested 70 years ago is still considered uninhabitable.

The detonation of an H-bomb is actually a series of explosions that take place in less than a second.

CHAPTER THREE

PHYSICAL OR CHEMICAL?

Everything that exists on Earth has characteristics that we use to identify it. For example, we may describe people in terms of eye or hair color or skin tone. Chemical elements are described in terms of their properties. Properties can be chemical or physical. A chemical change is one that changes the element into a different substance. For instance, when two hydrogen (H) atoms and one oxygen (O) atom bond, they become a different substance—water (H_2O). Oxidation is another example. For rust to form, iron (Fe), water, and oxygen must combine in a chemical reaction. Rust is a different substance from iron—it's brittle and flaky instead of hard and shiny.

Physical properties are things that can be observed without changing the element into a different element. This includes things like the element's phase at room temperature and its boiling and freezing points. For example, helium is still helium whether it's a gas or a liquid.

Elements exist in one of three physical states or phases—gas, liquid, or solid. Helium is a gas at room temperature. Room temperature is about 68°F (20°C). Mercury (Hg) is a liquid at room temperature. Gold (Au) exists as a solid at room temperature.

In its natural gaseous state, helium doesn't have a particular shape or size. A gas will expand to fill whatever container it's in. If there is no container, a gas will disperse into the atmosphere. Gases may also be stored in smaller containers under pressure.

Molecules in the solid phase of an element are fixed—a gold nugget at room temperature doesn't change shape. Molecules of a liquid flow. For example, mercury will change shape to fill the container it's in. Molecules of a gas are loosely arranged and will disperse if they're not contained.

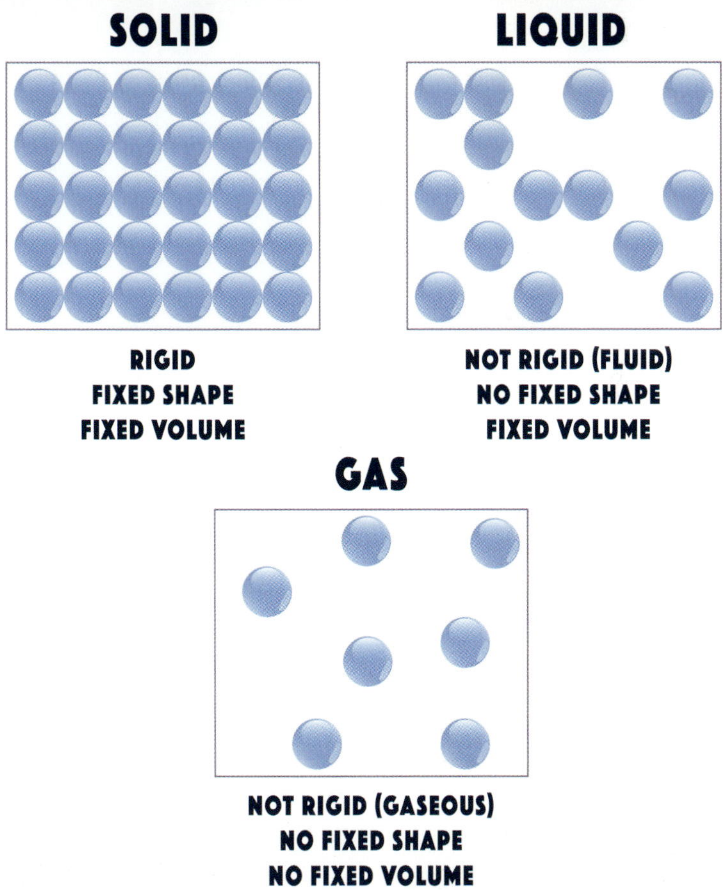

Helium gas is lighter than the mixture of gases that forms Earth's atmosphere. This is why helium balloons float.

Density is the mass of a substance contained in a specific volume. The density of a substance is found by dividing its mass by its volume. Each element has a unique density, and an element's density is useful in helping scientists identify it. Helium has a density of 0.0001664 gram/cubic centimeter at 68°F (20°C). The density of different gases is often compared to the density of air. If a gas has a lower density than air, an object containing it will float in air. In contrast, if a gas has a higher density than air, an object containing it will sink in air. If you have ever seen a helium balloon floating, you know that helium has a lower density than air's density under the same conditions.

At normal room temperature, helium is always a gas. However, if it is cooled to a low enough temperature or squeezed under enough pressure, helium can become a liquid or even a solid.

For helium to condense, or turn into a liquid, the temperature must reach a chilly -452.09°F (-268.94°C). This is the lowest condensation point of any known substance! In fact, this temperature is only 7.58°F (4.21 °C) higher than absolute zero (-459.67°F [-273.15°C]), the lowest temperature possible.

NOT JUST FOR FUN

Helium isn't just used to make balloons float. It serves a very important function in many medical and scientific facilities. Helium is one of the coldest substances on Earth. It's used to cool superconductors used for chemical research. It's also used to keep the magnets in magnetic resonance imaging (MRI) machines functioning. MRIs are critical for patient care, as they create high-resolution images of bones and organs—much clearer than X-ray images.

An MRI machine contains about 528.3 gallons (2,000 L) of liquid helium, which has to be replenished as it boils off. This is a concern to the medical community, as helium is a nonrenewable resource. There are known deposits in Algeria, Qatar, and Russia in addition to those found in the United States. However, the Federal Helium Reserve is in the process of being sold to a private company. Companies that produce MRI technology are researching new designs that can function with less helium.

Liquid helium is currently the coldest known substance on Earth.

Liquid helium is colorless. It's so transparent, or see-through, that it is impossible to see its surface with the naked eye. For helium to reach its freezing point (the temperature at which it becomes a solid), the temperature must drop even further to -457.87°F (-272.15°C). However, it is very difficult to make helium solidify. In fact, it is the only element that cannot be made into a solid at normal pressures simply by lowering its temperature. Scientists have only been able to solidify helium at extremely high pressures that are 25 times the normal atmospheric pressure.

SUPERCOOL!

When helium becomes a liquid at -452.09°F (-268.94°C), it is called helium I. Helium I exists at just a few degrees above absolute zero. Helium I naturally bubbles, but if it is cooled to -455.76°F (-270.98°C), just a few degrees above absolute zero, it becomes perfectly still and begins to expand. At these low temperatures, liquid helium is called helium II, a fluid that behaves very strangely. It behaves so differently from other fluids that scientists often refer to it as a superfluid.

Absolute zero is the temperature at which a thermodynamic system demonstrates the least amount of energy. Although molecules are never entirely still, at absolute zero they don't generate enough energy to translate to heat.

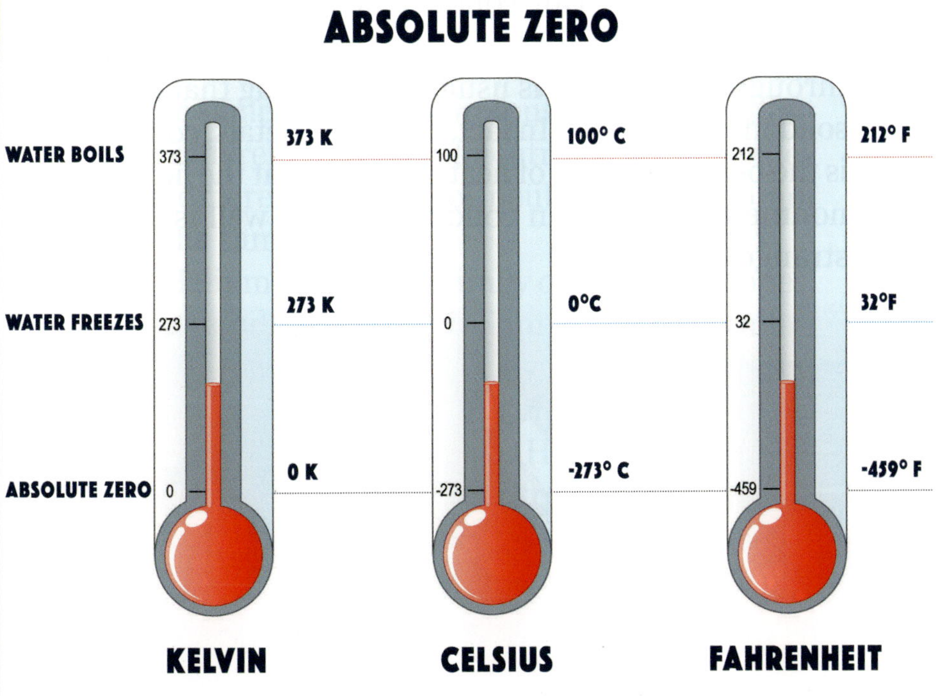

Solid helium can only be created in a lab. It requires extraordinarily low temperatures and high pressure to solidify.

Different isotopes of the same element have almost identical chemical properties. This means that they react in the same way. However, helium-3 has a lower boiling point than helium-4 and exhibits different properties when liquefied. Like helium-4, helium-3 can also act like a superfluid, but that occurs at temperatures lower than those for helium-4.

GETTING A REACTION

For a long time, scientists thought the noble gases, including helium, were totally inert or nonreactive. Helium atoms are very stable because their valence or outer electron shell is full. This means there's no space to share electrons with other atoms. It wasn't until 1962 that scientists were able to create reactions between noble gases and other elements in a lab. However, helium resisted efforts to combine it with other elements until 2017.

Scientists in Russia were able to combine helium atoms with sodium atoms to produce a compound (Na_2He) with a crystalline structure. They did this by subjecting the atoms to extremely high pressure, similar to that found at the earth's core. This created a different sort of bond—one that doesn't involve sharing electrons. The resulting compounds are stable.

Helium and sodium form a compound with molecules that look something like this familiar cube puzzle.

CHAPTER FOUR

WHERE DOES HELIUM COME FROM?

The United States is one of the world's largest producers of helium. In 2023 it produced about 2.8 billion cubic feet (79 million cubic meters). It's also one of the world's largest consumers of helium—consuming about 2 billion cubic feet (59 million cubic meters) in 2023. Helium is used to cool many kinds of high-tech equipment, including medical scanners and rocket engines. So, where does all this helium come from?

Helium is so light that Earth's gravity isn't strong enough to hold it in our atmosphere. Most of it drifts off into space, leaving only trace amounts in the air that we breathe. Most of the helium produced in the United States comes from underground. Helium is trapped in natural gas deposits deep in the earth's crust. Oil rigs must drill through layers of rock to reach these deposits. Then the helium must be "scrubbed" or separated from methane, nitrogen, and other elements found in natural gas. Finally a cryogenic separation process renders the helium into liquid form.

There are fewer than 20 liquid helium refineries in the world. Half are in the United States. The rest are found in Qatar, Russia, Algeria, and Australia—other places with large natural gas deposits.

It's not economically feasible to extract free-floating helium from the air, so we're dependent on underground gas fields. Researchers are exploring the possibility of additional deposits in Canada and Alaska.

Helium is extracted from natural gas deposits like this one.

NOBLE, INDEED

Sir William Ramsay was a British chemist. Although scientists had been aware of helium's existence since 1868, he's credited with being the first to liberate helium from a mineral sample (cleveite). Like Dmitri Mendeleev, Ramsay believed there were additional elements that would fit into gaps on the periodic table. He went on to discover argon, neon, krypton, and xenon. His research demonstrated that these inert gases, along with helium and radon, constituted their own special group on the periodic table—Group 18, the noble gases. He was awarded the Nobel Prize in Chemistry in 1904 for his discoveries.

Another noble gas was discovered by a German scientist named Friedrich Ernst Dorn (1848–1916) in 1900. It was named radon.

Nobel Prizes are awarded each year for outstanding achievements in literature, medicine, chemistry and physics. Perhaps the most famous is the related award for efforts toward peace. The awards were endowed by chemist Alfred Nobel, whose best-known invention is dynamite.

DRILLING FOR HELIUM

Drilling is used to locate natural gas reservoirs. The gas is then brought to the surface in pipes. The gas rises automatically because of the high pressure within the porous rocks where it is found. Once the gas has been extracted from the ground, it is usually transported in pipelines to a gas-processing plant. Here, helium can be isolated from the rest of the gaseous mixture.

Refineries contain a lot of pipes and other machinery.

Natural gas, which contains helium, must be drilled out of rock. This involves heavy tools like these special drill bits.

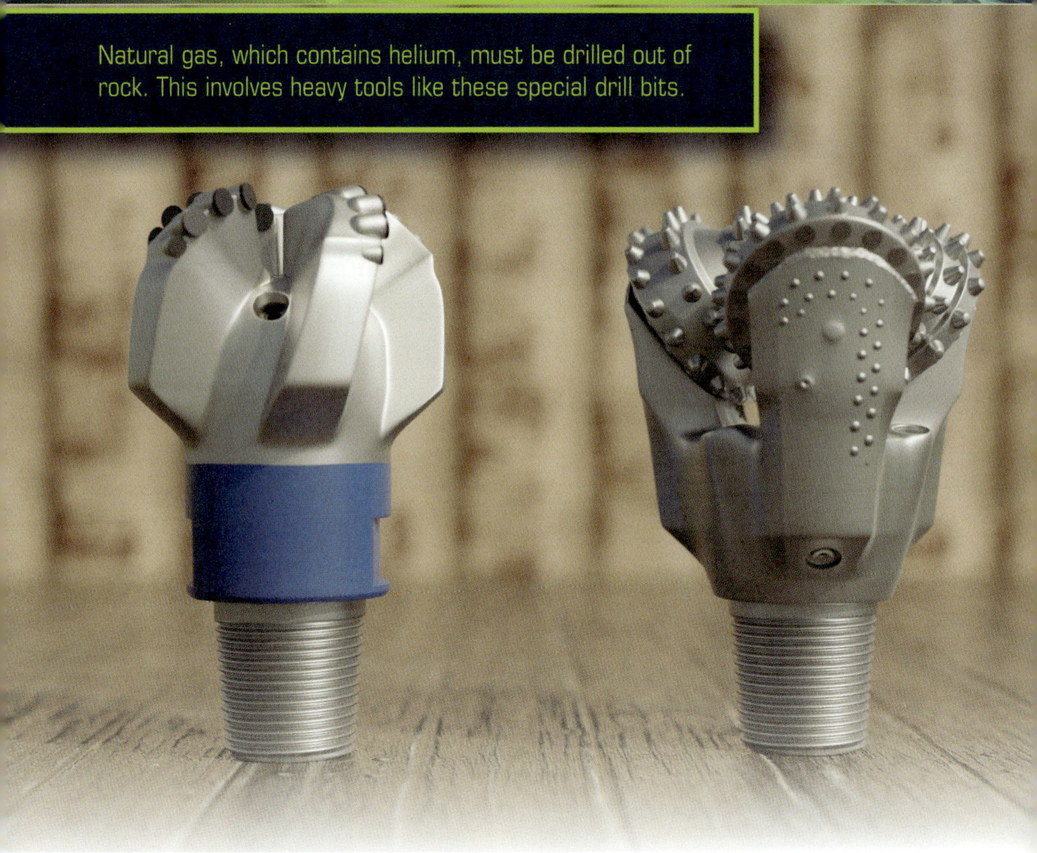

REFINING HELIUM

Helium is separated from methane and the other gases in natural gas by a process known as fractional distillation. Fractional distillation separates a mixture of liquids based on the boiling points of each element in the mixture. Every substance has a unique boiling point. As we have seen, helium's boiling point is -452.09°F (-268.94°C). Its boiling point is lower than that of all other substances. Because helium has such a low boiling point, it is very difficult to liquefy. Therefore, if natural gas is cooled, all the other gases will turn into liquids first, leaving only helium as a gas. The helium that remains is then pumped off.

Natural gas, once it has been restored to normal temperature, is sold to gas companies. The helium is shipped out under pressures of 1,800 to 2,000 pounds (816.5 to 907.2 kg) per square inch (6.5 sq cm) in small containers or special tank cars. This helium is used for many different purposes, as you will see in the following chapter. The United States is the leading producer of helium (it produces 80 percent), followed by Algeria, Russia, Canada, Poland, and Qatar. The helium that is produced is about 99.995 percent pure. That is a pretty remarkable achievement!

Once the gas has been drilled out of the earth, it's transported in pipelines. Natural gas must be refined before it can be used.

Inhaling helium can cause a person to lose consciousness very quickly, possibly resulting in a fall that causes additional injury.

WHAT HAPPENS WHEN YOU BREATHE HELIUM?

Everyone knows that if you inhale a little helium, you'll talk with a squeaky, high-pitched voice for a few minutes. That sounds like just a bit of silly fun, right? But what exactly happens when you breathe in helium? When you inhale helium, it replaces the oxygen in your lungs. While one breath probably won't hurt you, it's not a good thing to fool around with. Your body needs oxygen to function. When your lungs are full of helium instead, you'll feel dizzy. You might faint. If you breathe in too much helium, you could die.

WHAT IF WE RUN OUT OF HELIUM?

As we've discussed, helium is found on Earth deep underground in reservoirs of natural gas. To be used, natural gas must be drilled for and refined. You may have heard about fossil fuels and the search for sources of clean, renewable energy. Fossil fuels such as oil are considered nonrenewable resources because they take millennia to form from decomposed organic matter. Similarly, helium is formed by the radioactive decay of elements like uranium. This process takes thousands of years. In addition, helium is so light that it easily escapes from the atmosphere into space.

Some scientists are concerned that we may use up all the available helium before more can form. Others say that Earth will continue to produce helium for a long time. Since it's critical to medicine and science, researchers are trying to design new technology that can operate using less helium or by using other elements such as hydrogen. Companies are also looking for new desposits all the time.

Many of the oil deposits being drilled today were formed more than 66 million years ago.

CHAPTER FIVE

HOW WE USE HELIUM

Initially, scientists didn't think helium did much of anything. That's why it was classified as inert. However, today we use helium for many different purposes, including entertainment, science, and heavy industry.

Of course, you're probably familiar with helium-filled party balloons, but did you know that the airbags that protect you in a car crash are also filled with helium? Helium is used in the manufacture of the computer chips that power your personal electronics. It's also used for high-speed internet and cable TV.

Helium is used in arc welding and other industries to protect metals from oxidation. Because it's mostly inert, helium doesn't react with other elements to create corrosion or trigger explosions. For this reason, it's also used to clean rocket fuel tanks. Helium is the preferred gas for filling weather balloons because it's nonflammable.

For a substance people initially thought was useless, helium sure gets around! It's used to pressurize aircraft and to treat breathing disorders like asthma. Helium also makes up part of the air in scuba tanks.

A mixture of helium and oxygen called heliox is sometimes used in emergency medicine. Because helium is lighter than nitrogen, it takes less work for an injured body to breathe it in.

A swimmer without scuba gear can go about 20 feet (6.09 m) deep.

Have you ever swum all the way to the bottom of the deep end of a pool? You might have noticed pressure in your ears. This is because water is heavier than air. When you swim underwater, the weight of the water presses on your body. When people swim very deep underwater, they use an air tank to breathe. The tank holds the air at the same pressure as the water, so the diver's lungs don't collapse.

However, when high-pressure gases come in contact with water, they dissolve into the water. This is how carbonated beverages, such as soda, are made. To make carbonated water, the water is exposed to high-pressure carbon dioxide gas. This forces the gas to dissolve in the water. In the same way, a diver who stays in deep water for a while will have dissolved nitrogen and oxygen gases from the air tank forced into their bloodstream.

Excess nitrogen and oxygen in the blood can make a diver sick. Nitrogen narcosis affects judgement and can be similar to drinking alcohol—not a good thing when you're underwater. Although oxygen is necessary for human life, too much in the bloodstream can cause convulsions.

Decompression sickness (or "the bends") occurs when a diver surfaces too quickly. A diver needs to ascend slowly so their body can adjust to the change in pressure. Otherwise, nitrogen gas forms bubbles in their bloodstream.

Scuba tanks make it possible for people to dive safely to much greater depths than they can unassisted.

Helium can help divers avoid dangerous problems such as nitrogen narcosis and oxygen toxicity. Helium is not very soluble in water or blood. This means that not a lot of helium will enter a diver's bloodstream, even under high pressure. For this reason, in many gases used for breathing, part of the nitrogen and oxygen is replaced by helium. Helium is easy to breathe at high pressures and, unlike oxygen and nitrogen, it does not become toxic. Some common helium-containing breathing gases are trimix, a mixture of helium, oxygen, and nitrogen; heliox, which contains helium and oxygen; and hydreliox, a mixture of hydrogen, helium, and oxygen. These helium-containing breathing gases can help save a diver's life, but they can also make that diver sound a little bit like a cartoon character!

PRESERVING PRICELESS ARTIFACTS

Two of the United States' most important documents are the Declaration of Independence and the Constitution. The Declaration of Independence marked the first official step toward separating the 13 colonies from the control of Great Britain—and the first step toward creating the United States of America. The document was mainly written by Thomas Jefferson (1743–1826). Congress voted to adopt it July 4, 1776. This is why the Fourth of July is celebrated as Independence Day today. Following the American Revolution (1775–1783), in which Americans fought the British to win their independence, the Founding Fathers crafted the Constitution. It was accepted by all the states on May 29, 1790. The Constitution is the basis for the government of the United States. It explains how the government is formed, who makes it up, and how laws are to be passed and amended.

For half a century, helium protected some of the most important—and fragile—documents in America.

Noble gases are used to preserve priceless historic documents in the National Archives.

The Declaration of Independence and the Constitution were described by the poet and Librarian of Congress Archibald MacLeish (1892–1982) as "fragile objects which bear so great a weight of meaning to our people." Indeed, these priceless documents are quite fragile. They are handwritten with iron-gall ink on parchment. If left out in the air, atmospheric gases, such as oxygen, would surely destroy them. For this reason, it was decided that the documents would be enclosed in a case with the least reactive gas. What better than helium? In 1952, the frames holding the Declaration of Independence and the Constitution were emptied of air and then filled with 1 percent water vapor and 99 percent helium and placed in the National Archives in Washington, D.C. What an honor for helium to be entrusted with such important documents!

Due to deterioration of the glass that enclosed the documents, they were transferred to new containers made of aluminum (Al), titanium (Ti), and glass in 2001. The enclosures that now surround the documents are filled with argon, another of the noble gases. Argon atoms are larger than helium atoms and are less likely to leak out. However, for nearly 50 years, helium did its job well.

WELDING

Helium is very useful when it comes to welding. Welding is a process in which two or more pieces of metal are melted and joined together. Welding is used in the construction of bridges, ships, cars, airplanes, and other objects. Helium is used for both arc welding and laser welding. In arc welding, the heat that joins the metals together comes from electricity. Laser welding uses a laser beam to generate heat. Welding can be very dangerous. When the metals get very hot, they can react with oxygen in the air and start to burn. Oxygen can also ruin

There are different grades of helium gas. This is based on their purity. Some uses of helium call for a purer form of this gas than others.

the metals that welders are trying to join. For this reason, helium is used to protect the metal. Helium does this by "shielding" the metals from oxygen in the atmosphere. Helium is used for the welding of metals like stainless steel, aluminum, copper, nickel (Ni), titanium, and magnesium (Mg). As these metals are welded, helium gas is continuously blown over the area. Helium pushes air away from the metals, keeping oxygen and moisture from attacking them. Since helium is unreactive, it does not react with the air or the metals. Helium is also used in this process with other shielding gases, such as argon.

Welding requires special safety procedures to protect the welder, their project, and their work environment. Helium protects metal from oxidation. Because it's not flammable, it helps prevent sparks from igniting.

Many people know of the showstopping balloons in the Macy's Thanksgiving Day Parade. Each year the parade uses about 400,000 to 700,000 cubic feet (11,326.7 to 19,821.8 cubic meters) of helium to fill its balloons.

LIGHTER THAN AIR

Helium balloons are used for everything from scientific research to military observation. However, everyone's favorite helium balloons are probably the colorful ones used for fun. Helium gained its fame as a lifting gas. It is lighter than air, so it floats on air in the same way that a plastic bottle filled with air floats on water.

Any gas that's less dense or lighter than air can function as a lifting gas. This includes:

- hydrogen (H)
- nitrogen (N)
- methane (CH_4)
- ammonia (NH_3)
- neon (Ne)
- hot air
- natural gas
- water vapor

However, many of them have drawbacks. Hydrogen is responsible for airship disasters due to its explosive nature. Methane and natural gas are also highly flammable. Ammonia is flammable in addition to being a poison that irritates the mucous membranes and lungs. Hot air is used primarily for recreational hot-air balloons.

For a gas to see widespread commercial use as a lifting gas, it must be plentiful, economical to obtain and process, nontoxic, and nonflammable. Imagine filling party balloons with a gas that could catch fire, or make people sick if they breathed it? Although there are concerns about future supplies, currently helium is the best fit.

Helium is found everywhere from scientific facilities to industrial sites to the balloon tank in your local party store. It fills weather balloons and party balloons. It serves as a coolant for supercomputers and medical scanners. It helps divers explore the bottom of the sea and astronauts explore space. It makes our lives safer and a little more fun. That's quite a résumé for a gas that scientists once thought served no practical purpose.

THE PERIODIC TABLE OF ELEMENTS

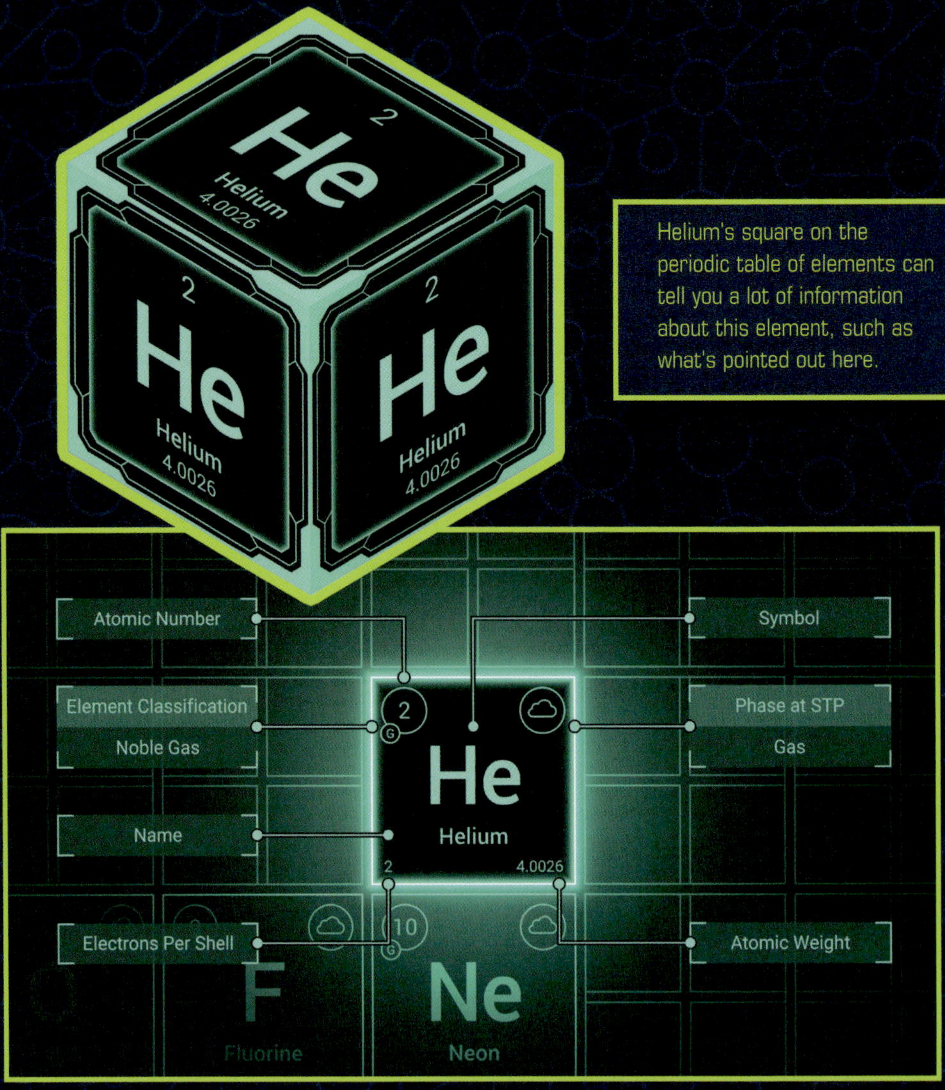

Helium's square on the periodic table of elements can tell you a lot of information about this element, such as what's pointed out here.

PERIODIC TABLE OF ELEMENTS

- Alkali metals
- Alkaline earth metals
- Transition metals
- Metal
- Metalloids
- Nonmetals
- Halogen
- Noble gases
- Lanthanoids
- Actinoids

1 H Hydrogen 1.008																	2 He Helium 4.0026
3 Li Lithium 6.94	4 Be Beryllium 9.0122											5 B Boron 10.81	6 C Carbon 12.011	7 N Nitrogen 14.007	8 O Oxygen 15.999	9 F Fluorine 18.998	10 Ne Neon 20.180
11 Na Sodium 22.990	12 Mg Magnesium 24.305											13 Al Aluminium 26.982	14 Si Silicon 28.085	15 P Phosphorus 30.974	16 S Sulfur 32.06	17 Cl Chlorine 35.45	18 Ar Argon 39.948
19 K Potassium 39.098	20 Ca Calcium 40.078	21 Sc Scandium 44.956	22 Ti Titanium 47.867	23 V Vanadium 50.942	24 Cr Chromium 51.996	25 Mn Manganese 54.938	26 Fe Iron 55.845	27 Co Cobalt 58.933	28 Ni Nickel 58.693	29 Cu Copper 63.546	30 Zn Zinc 65.38	31 Ga Gallium 69.723	32 Ge Germanium 72.630	33 As Arsenic 74.922	34 Se Selenium 78.971	35 Br Bromine 79.904	36 Kr Krypton 83.798
37 Rb Rubidium 85.468	38 Sr Strontium 87.62	39 Y Yttrium 88.906	40 Zr Zirconium 91.224	41 Nb Niobium 92.906	42 Mo Molybdenum 95.95	43 Tc Technetium (98)	44 Ru Ruthenium 101.07	45 Rh Rhodium 102.91	46 Pd Palladium 106.42	47 Ag Silver 107.87	48 Cd Cadmium 112.41	49 In Indium 114.82	50 Sn Tin 118.71	51 Sb Antimony 121.76	52 Te Tellurium 127.60	53 I Iodine 126.90	54 Xe Xenon 131.29
55 Cs Caesium 132.91	56 Ba Barium 137.33	57-71	72 Hf Hafnium 178.49	73 Ta Tantalum 180.95	74 W Tungsten 183.84	75 Re Rhenium 186.21	76 Os Osmium 190.23	77 Ir Iridium 192.22	78 Pt Platinum 195.08	79 Au Gold 196.97	80 Hg Mercury 200.59	81 Tl Thallium 204.38	82 Pb Lead 207.2	83 Bi Bismuth 208.98	84 Po Polonium (209)	85 At Astatine (210)	86 Rn Radon (222)
87 Fr Francium (223)	88 Ra Radium (226)	89-103	104 Rf Rutherfordium (267)	105 Db Dubnium (268)	106 Sg Seaborgium (269)	107 Bh Bohrium (270)	108 Hs Hassium (277)	109 Mt Meitnerium (278)	110 Ds Darmstadtium (281)	111 Rg Roentgenium (282)	112 Cn Copernicium (285)	113 Nh Nihonium (286)	114 Fl Flerovium (289)	115 Mc Moscovium (290)	116 Lv Livermorium (293)	117 Ts Tennessine (294)	118 Og Oganesson (294)
			57 La Lanthanum 138.91	58 Ce Cerium 140.12	59 Pr Praseodymium 140.91	60 Nd Neodymium 144.24	61 Pm Promethium (145)	62 Sm Samarium 150.36	63 Eu Europium 151.96	64 Gd Gadolinium 157.25	65 Tb Terbium 158.93	66 Dy Dysprosium 162.50	67 Ho Holmium 164.93	68 Er Erbium 167.26	69 Tm Thulium 168.93	70 Yb Ytterbium 173.05	71 Lu Lutetium 174.97
			89 Ac Actinium (227)	90 Th Thorium 232.04	91 Pa Protactinium 231.04	92 U Uranium 238.03	93 Np Neptunium (237)	94 Pu Plutonium (244)	95 Am Americium (243)	96 Cm Curium (247)	97 Bk Berkelium (247)	98 Cf Californium (251)	99 Es Einsteinium (252)	100 Fm Fermium (257)	101 Md Mendelevium (258)	102 No Nobelium (259)	103 Lr Lawrencium (266)

- C Solid
- Hg Liquid
- H Gas
- Rf Unknown

GLOSSARY

atmosphere: The air that surrounds Earth.
atom: The smallest part of an element having the chemical properties of that element.
atomic weight: The average mass (in units) of the atoms of the isotopes of an element.
bond: An attractive force that links two atoms together.
chemical change/reaction: A change in which one kind of matter is turned into another kind of matter.
compound: A substance made up of two or more elements that have combined chemically.
cryogenic: Of or relating to the production of very low temperatures.
crystalline: Resembling crystal.
electron: A negatively charged particle found outside of the nucleus of an atom.
element: A substance made of only one type of atom.
energy: The ability to do work (or produce heat).
gravity: A force of attraction between masses.
mass: A measure of the amount of matter in something.
matter: What things are made of. Matter takes up space and has mass.
nucleus: The positively charged central portion of an atom.
period: A horizontal row in the periodic table.
phase: One of the three fundamental states of matter, which are solid, liquid, and gas.
pressure: The force acting on a surface.

proton: A positively charged particle within the nucleus of an atom. The number of protons and electrons are almost always equal in an atom.
room temperature: The average indoor temperature at which experiments are performed. Scientists consider room temperature to be approximately 68°F (20°C).
thermodynamics: The branch of physics that deals with the mechanical action or relations of heat.
valence shell: The outermost shell of an atom containing the valence electrons.
viscosity: A fluid's resistance to flow.
volume: The amount of space that something occupies.

FOR MORE INFORMATION

AMERICAN CHEMICAL SOCIETY

1155 16th Street NW
Washington, DC 20036
(800) 227-5558
Website: www.acs.org
Facebook: @AmericanChemicalSociety
Instagram and X: @AmerChemSociety
ACS is a global professional community of chemists.

CHEMICAL INSTITUTE OF CANADA (CIC)

90-2420 Bank Street
Ottawa, Ontario, Canada
K1V 8S1
(613) 232-6252
Website: www.cheminst.ca/
X: @CIC_ChemInst
The CIC facilitates the research and cooperation of chemists and chemical engineers in Canada.

COMPRESSED GAS ASSOCIATION (CGA)

8484 Westpark Drive, Suite 220
McLean, Virginia 22102
703-788-2700
Website: www.cganet.com
X: @CGA_Safety
The CGA is a trade association dedicated to creating educational materials and developing safe handling and workplace regulations for industries that utilize compressed gases.

INTERNATIONAL UNION OF PURE AND APPLIED CHEMISTRY (IUPAC)

IUPAC Secretariat
79 T.W. Alexander Drive
Research Commons Building 4501, Suite 190
Research Triangle Park, NC 27709, USA
(919) 485 8700
Website: iupac.org/
Facebook: @iupac.org
X: @iupac
IUPAC is the global chemistry organization that maintains the periodic table of elements.

INTERNATIONAL UNION OF PURE AND APPLIED PHYSICS (IUPAP)

Fondazione Internazionale Trieste per il Progresso e la Libertà delle Scienze
c/o ICTP
Strada Costiera 11
34151 Trieste
Website: iupap.org/
Facebook: @iupap
X: @iupap_physics
IUPAP's mission is to assist in the global advancement of physics and facilitate international scientific cooperation.

THE LIGHTER-THAN-AIR SOCIETY

526 S. Main Street, 406
Akron, OH 44311
Web site: www.blimpinfo.com
The Lighter-Than-Air Society is a nonprofit group dedicated to the study of all forms of buoyant aircraft.

FOR FURTHER READING

Congdon, Lisa. *The Illustrated Encyclopedia of the Elements: The Powers, Uses, and Histories of Every Atom in the Universe*. San Francisco, CA: Chronicle Books, LLC, 2021.

DK. *Super Simple Chemistry: The Ultimate Bitesize Study Guide*. New York, NY: DK Publishing, 2020.

McHenry, Ellen Johnston. *The Chemical Elements Coloring & Activity Book: A Fun and Interactive Guide to the World of Chemistry*. Independently Published, 2021.

Silver, Donald, and Patricia Wynne. *My First Book About Chemistry*. Mineola, NY: Dover Publications, 2020.

Thomas, Isabel, and Sara Gillingham. *Exploring the Elements: A Complete Guide to the Periodic Table*. New York, NY: Phaidon Press, Inc., 2021.

Zovinka, Edward P., and Rose A. Clark. *A Kids' Guide to the Periodic Table: Everything You Need to Know About the Elements*. Emeryville, CA: Rockridge Press, 2020.

INDEX

ABOUT THE AUTHOR

KATHLEEN A. KLATTE is the author of many nonfiction books for children and teens. Topics she has written about range from animals and nature to constitutional law to unusual career choices. She lives in New York with one cat and far too many books and Legos.

PHOTO CREDITS

Cover, p. 1 remotevfx.com/Shutterstock.com; Cover, pp. 1, 3-64 pluie_r/Shutterstock.com; Cover, pp. 1, 3-64 Oksancia/Shuttestock.com; p. 5 Demkat/Shutterstock.com; p. 7 grayjay/Shutterstock.com; p. 9 buradaki/Shutterstock.com; p. 10 Ipajoel/Shutterstock.com; p. 11 Srikanth_K/Shutterstock.com; p. 12 Anton Veselov/Shutterstock.com; p. 13 vchal/Shutterstock.com; p. 15 Watthana Tirahimonchan/Shutterstock.com; p. 18 TonelloPhotography/Shutterstock.com; p. 19 New Africa/Shutterstock.com; p. 21 Jag_cz/Shutterstock.com; p. 23 Alones/Shutterstock.com; p. 25 Anshuman Rath/Shutterstock.com; p. 26 Selena3726/Shutterstock.com; p. 28 kittirat roekburi/Shutterstock.com; p. 29 Designua/Shutterstock.com; p. 30 Ziadi Lotfi/Shutterstock.com; p. 31 BD ARMANS/Shutterstock.com; p. 32 tunasalmon/Shutterstock.com; p. 33 Kristof Desi/Shutterstock.com; p. 35 Maurizio Bersanelli/Shutterstock.com; p. 36 Bertil Jonsson/Shutterstock.com; p. 37 Rudmer Zwerver/Shutterstock.com; p. 38 Paradigm Studio/Shutterstock.com; p. 39 Mike Mareen/Shutterstock.com; p. 40 MediaClemm/Shutterstock.com; p. 41 Maksim Safaniuk/Shutterstock.com; p. 43 Davizro Photography/Shutterstock.com; p. 44 Dulyanut Swdp/Shutterstock.com; p. 45 Sun_Shine/Shutterstock.com; p. 46 Sven Hansche/Shutterstock.com; p. 47 PT Hamilton/Shutterstock.com; p. 48 Jaclyn Vernace/Shutterstock.com; p. 50 Andrewshots/Shutterstock.com; p. 51 MexChriss/Shutterstock.com; p. 52 gary718/Shutterstock.com; p. 54 ShannonChocolate/Shutterstock.com; p. 55 View_From_My_Lens/Shutterstock.com.